Do You Like to Bike?

by Fawn Bailey

Riding a bike is a fun thing to do. It takes lots of force to pedal one, too.

There are all kinds of bikes that people can ride. Which one do you like? It's time to decide!

Road bikes are made
for riding with speed.
They have smooth tires,
and that's what you'll need.

Pull on your helmet.
Snap it under your chin.
Push on the pedals,
and the wheels will spin.

The force of gravity
pulls you down.
Push hard to get
to higher ground.

Mountain bikes are good for gripping. The thick, bumpy tires keep you from slipping.

A bicycle for two
can be easy to learn.
You both have to lean
when you're ready to turn.

The work is shared
when two ride as one.
That's why tandem bikes
can be so much fun!

Which bike is the best?
Here is a clue.
If you like rocky trails,
a mountain bike is for you.

Do you like to go fast
on a bike that can glide?
A road bike is sure
to give a smooth ride.

It's important to think about motion and speed. So be sure to get the bike that you need.